Hla Myo Tun
Zaw Min Naing
Win Khine Moe

Time-Reversal Focusing for Human Computer Interface Application

Hla Myo Tun
Zaw Min Naing
Win Khine Moe

Time-Reversal Focusing for Human Computer Interface Application

LAP LAMBERT Academic Publishing

Impressum / Imprint

Bibliografische Information der Deutschen Nationalbibliothek: Die Deutsche Nationalbibliothek verzeichnet diese Publikation in der Deutschen Nationalbibliografie; detaillierte bibliografische Daten sind im Internet über http://dnb.d-nb.de abrufbar.
Alle in diesem Buch genannten Marken und Produktnamen unterliegen warenzeichen-, marken- oder patentrechtlichem Schutz bzw. sind Warenzeichen oder eingetragene Warenzeichen der jeweiligen Inhaber. Die Wiedergabe von Marken, Produktnamen, Gebrauchsnamen, Handelsnamen, Warenbezeichnungen u.s.w. in diesem Werk berechtigt auch ohne besondere Kennzeichnung nicht zu der Annahme, dass solche Namen im Sinne der Warenzeichen- und Markenschutzgesetzgebung als frei zu betrachten wären und daher von jedermann benutzt werden dürften.

Bibliographic information published by the Deutsche Nationalbibliothek: The Deutsche Nationalbibliothek lists this publication in the Deutsche Nationalbibliografie; detailed bibliographic data are available in the Internet at http://dnb.d-nb.de.
Any brand names and product names mentioned in this book are subject to trademark, brand or patent protection and are trademarks or registered trademarks of their respective holders. The use of brand names, product names, common names, trade names, product descriptions etc. even without a particular marking in this work is in no way to be construed to mean that such names may be regarded as unrestricted in respect of trademark and brand protection legislation and could thus be used by anyone.

Coverbild / Cover image: www.ingimage.com

Verlag / Publisher:
LAP LAMBERT Academic Publishing
ist ein Imprint der / is a trademark of
OmniScriptum GmbH & Co. KG
Heinrich-Böcking-Str. 6-8, 66121 Saarbrücken, Deutschland / Germany
Email: info@lap-publishing.com

Herstellung: siehe letzte Seite /
Printed at: see last page
ISBN: 978-3-659-81162-3

Zugl. / Approved by: New Delhi, Indian Institute of Technology, Delhi, Postdoctoral Research Report, 2013

Dedication

To Myat Su Nwe with Love

Acknowledgments

This research report is made possible through the help and support from everyone, including: parents, teachers, family, friends, and in essence, all sentient beings. Especially, please allow me to dedicate my acknowledgment of gratitude toward the following significant advisors and contributors:

First and foremost, I would like to thank **Professor Dr. Arum Kumar** (Professor and Head, Centre for Applied Research in Electronics, Indian Institute of Technology, Delhi) for his most support and encouragement. He kindly read my research report and offered invaluable detailed advices on grammar, organization, and the theme of the research report.

A special gratitude I give to **Dr. Rajiv Kumar**, Scientist 'D', International Division, Department of Science and Technology, Ministry of Science and Technology, whose contribution in stimulating suggestions and encouragement, helped me to coordinate my project especially for doing this research in Indian Institute of Technology, Delhi (IITD).

Third, I would like to express gratitude Department of Science and Technology (DST), Government of India for supporting the research grant of P C Ray Senior Fellowship award.

Finally, I sincerely thank to my parents, family, and friends, who provide the advice and financial support. The product of this research report would not be possible without all of them.

Abstract

In this research report, time-reversal focusing was designed for human computer interface applications by exploiting the ultra-wideband (UWB) acoustics waves. In contrast, conventional inverse scattering techniques are restricted by the diffraction limit on their ability to determine geometrical features. Additionally, frequency-domain inverse techniques are affected from the particular properties of the medium for each diverse realization. UWB signals that can be used to get better detection capabilities combat a number of scattering problems. They can pilot to statistically stable imaging techniques that depend only on the statistical properties of the random medium. The scheme is to focus array of transducers from any direction by using Time Reversal-Ultra Wideband (TR-UWB).

In this work the time reversal focusing based on finite difference time domain (FDTD) method to localize source by using MATLAB GUI. There are three analyses of free space model, waveguide model and user define model for source location in a perfectly matched layer boundary. We have shown the energy spectrum for propagation of acoustics wave in a room to focus by using array of transducers and source. In these plots we first transmit the signals from source to the array of transducers and then the signals can be retransmitted back from array of transducers to the source by using time reverse algorithm. The reverse signals can be focused the location of source based on the proposed method. The simulation results are evaluated to localize the source by using FDTD method in this work. The application areas of this work are mentioned in this work.

2

Table of Contents

Pages

Acknowledgements   1

Abstract   2

Table of Contents   3

Chapter 1   Introduction   5

    1.1   Research Motivation   5

    1.2   Research Objectives   6

    1.3   Research Direction   7

    1.4   Contributing Research Areas   7

    1.5   Originality of the Report   7

    1.6   Organization of the Report   7

Chapter 2   Literature Review   9

    2.1   Introduction   9

    2.2   What is Human Computer Interface Technology (HCI)?   9

    2.3   Previous Investigation on Time Reversal Mirror (TMR)   9

    2.4   Finite Difference Time Domain Method   11

    2.5   Proposed Model   14

    2.6   Room Impulse Responses Model Development 15

Chapter 3   Implementation and Simulation Results   17

    3.1   Overall System Flowchart   18

    3.2   Software Development   18

        3.2.1   Specified Boundary   19

        3.2.2   Specified Source   20

        3.2.3   Specified Transducer Array 20

        3.2.4   Specified Medium   20

        3.2.5   Choice of Modulation Method 21

        3.2.6   Transmit Signal   21

	3.2.7 Time Reversal Process	21
	3.2.8 Re-transmit Signal	21
	3.2.9 Localized Source	23
3.3	Development of Room Impulse Response	23
3.4	Flowchart of FDTD Scheme	24
3.5	Analysis of Source Localization for Proposed Research Work	24
3.6	Analysis of Bit Error Rate	33
3.7	Simulation Results	33
Chapter 4	Application Areas of Proposed Research	45
Chapter 5	Conclusion	46
References		47

Chapter 1

Introduction

1.1 Research Motivation

The advancement of human-computer interfaces (HCIs) has become progressively more popular. This is not surprising due to the propagation of user demands ranging from internet applications to software requirements that engage complex user-defined input functions. Coupled with the increasing need for ease of use for such systems, research has since been determined on developing new HCIs. Existing HCIs such as the alphanumeric keyboard and optical pointing device present physical limitations. In addition, touch interfaces (TIs) have been urbanized to deal with these limitations by allowing users to operate the software through the use of finger manoeuvre. These TIs have since been integrated to the personal computer for cyber world applications including virtual/online gaming.

In this research work, we propose to address this cost limitation by utilizing a network of low cost surface mounted accelerometers. When mounted on solid surfaces, these sensors have the benefit of adapting an ordinary surface into a touch surface. Consequently, the aim of this signal processing module is to localize a finger tap based on signals received from these sensors. Existing algorithms such as that proposed in [1] necessitate training of the system prior to usage; the signals at each target location will be recognized through a matching process during usage. This presents an enormous disadvantage since a significant number of touch locations have to be trained for each location prior to usage. On a technical level, a large number of touch locations will frequently translate to a significant amount of memory and processing requirements.

Source localization involves a number of sensors that are spatially separated. The localization accuracy is reliant not only on the quality of the measurements, the number of sensors and the inspection period, but also on how the sensors are being arranged. This is the geometric consequence that is often termed as geometric dilution of precision [2] in geo-location literature. It is therefore important to explore the sensor arrangement that can accomplish the best localization accuracy. The objective of source localization is to recognize the location of an emitting source based on the signal measurements from a number of sensors. In active localization, the source signal waveform and possibly its starting time are identified to the sensors, so that the time of arrival (TOA) of the signal to the sensors can be extracted for source localization. Possibly the more challenging scenario is passive localization, where the source signal and its time-stamp are not identified to the receivers. Moreover, the signal-to-noise ratio (SNR) in the sensor measurements could be very low and a long observation time is needed to attain the source location. Passive localization frequently utilizes time difference of arrival (TDOA) or angle of arrival (AOA) of the source signal to different sensors. Source localization has established numerous applications nowadays in radar, sonar, wireless communications, and sensor networks [3]–[5].

Time reversal is a technique in which a signal is pre-filtered such that it focuses both in time and space. Time Reversal Mirror (TRM) technique, in the virtue of its high resolution in the heterogeneous media, has been widely applied in the area of acoustics and electromagnetics. In this system, FDTD algorithm is employed to simulate the acoustic wave propagation in the UWB environment. A number of tests with numerical data to validate the 2-D acoustic TRM imaging in the context of UWB were investigated and discussed. There are two kinds of stimulation of physical objects: passive and active modes. In the passive mode any change in the acoustic properties of an object, due to its vibration as a consequence of interaction (knocking, tapping etc.), is detected and then used to estimate the location of the interaction. In the active mode, the absorption of acoustic energy at the contact point of an object surface must be ascertained. This investigation focused on signal processing algorithm development, acoustic wave propagation analysis and simulation.

1.2 Research Objectives

- To develop the time-reversal focusing for human computer interface applications by using UWB acoustics waves

- To design the Graphical User Interface (GUI) based on time-reversal focusing with the help of MATLAB

1.3 Research Direction

In this research work, single transmit multiple receive sensors in the acoustics domain with wideband signals will be focused. The acoustics simulation is performed using the FDTD algorithm with PML boundary conditions.

- Theoretical analysis of the time reversal based algorithms
- Simulations that tested the time reversal algorithms in a number of realistic scenarios;

1.4 Contributing Research Areas

While the main contribution of this research report is the development of time-reversal focusing for human computer interface applications by using UWB acoustics waves, a number of the other contributions can be described as follows:

- An investigation into the suitability of source localization for HCI development
- The development of methods to localized of senor arrays for HCI deployment

1.5 Originality of the Report

The originalities of the report are:

- Finding the evidence of time reversal focusing for HCI applications
- An implementation of MATLAB GUI application and evaluation performance

1.6 Organization of the Report

This research report is organized into six chapters. Chapter 2 discussed the conceptual framework of HCI and reviews current state of the art implementations. Chapter 2 also gives some background information regarding the source localization. Various functional time reversal mirrors that have been developed throughout the years are also compared, and time reversal focusing approach is introduced.

7

Chapter 3 focused on the development of time-reversal focusing for human computer interface applications by using UWB acoustics waves with the help of MATLAB GUI programming. The simulation results on the development of time-reversal focusing for human computer interface applications by using UWB acoustics waves have been described in this cahpter.

Chapter 4 mentions the application areas of the proposed research work were also described in this chapter.

Chapter 5 discusses the research outcomes and expresses the future work on the proposed research work.

Chapter 2

Literature Review

2.1 Introduction

Approaches for human-machine interface rely on the need for keyboard and the mouse. As new software applications continue to evolve, one of the main drawbacks of such input devices is that they impede ease of operating software or manipulating data which require complex user input operations. As a result, these devices limit the scope and functionality of the PC. This project aims to develop a new paradigm by transforming everyday objects such as tabletops and glass panels into a human-machine interface using a network of low-cost surface mounted sensors. Signal processing algorithms will be developed to localize and track movement of or tapping of fingers on different materials. These locations can be used to control software applications.

2.2 What is Human Computer Interface Technology (HCI)?

"The goal of improving the Human and Computer Interface (HCI) is to make our work with the computer more "natural." It is to capitalize upon our sensory network - sight, hearing, speech, touch, taste and smell, and to eliminate the "crude" devices with which we must now deal - the keyboard, mouse, and computer monitor."

2.3 Previous Investigation on Time Reversal Mirror (TMR)

Amir Sulaiman et al [1], proposed a new paradigm of touch interface that allows one to convert daily objects to a touch pad through the use of surface mounted sensors. To achieve a successful touch interface, localization of the finger tap is important. An inter-disciplinary approach to improve source localization on solids by means of a mathematical model was presented. It utilizes mechanical vibration theories to simulate the output signals derived from sensors mounted on a physical surface. Utilizing this model, an insight into how phase is distorted in vibrational waves within an aluminium plate which in turn serves as a motivation for his work was provided. A source localization algorithm based on the phase information of the received signals was proposed. The performance of our algorithm using both simulated and recorded data was verified.

Tarik Yardibi et al [6] have presented a nonparametric and hyper parameter, free-weighted, least squares-based iterative adaptive approach for amplitude and phase estimation (IAA-APES) in array processing. IAA-APES is a nonparametric, hyper parameter free algorithm that is designed to work under severe snapshot limitations and for uncorrelated, partially correlated, and coherent sources, as well as for arrays with arbitrary geometries. Because of the similarities between many active sensing applications and passive array processing, IAA-APES can be applied to these cases as well, without any essential modifications.

Rudolf Sprik [7] has suggested a time reversed reconstruction of the original signal can be obtained by recording the evolving signal, then reversing the recorded signal, and sending the reversed signal again through the system. The reconstruction of the time reversed signal occurs in time as well as in space. The process is very robust and works even in systems with multiple scattering of the waves due to complex boundaries or heterogeneous material properties. Examples of such complex signals are the 'Coda' response in seismology and reverberation in room acoustics. In open systems where the waves can escape the region of interest, multiple transducers at the boundary of the region are required to perform an accurate reconstruction. The array of transducers acts as a time-reversal mirror refocusing the reversed signal to the original location of the source. The time reversal process is very robust and works with a relatively small arrays covering only part of the boundary, limited section of the recorded signal, or even 1-bit digitalization of the recorded signal.

Xueli Sheng et al [8] have discussed passive selective focusing and localization methods based on acoustic vector sensors array's iterative TRM technique is studied and dummy iterative time reversal mirror (DITRM). Multi-passive targets detecting in multi-path and environmental noise conditions is one of the main functions for sonar signal processing. The output of acoustic vector sensors array DITRM can not only refocuses the incident acoustic field back to the origin of a probe signal

10

regardless of the complexity of the medium, but the arrays of transducer focuses the source on the perfectly matched layer. The validity of that method was verified in the simulation experiments. The results show that this technology has higher performance and precision because of TRM vector array processing, and is more robust than pressure only DITRM.

Blaine M. Harker et al [9] have described time reversal (TR) utilizes an array of transducers, a time reversal mirror (TRM), to locate sources. TR is applied to simple sources using steady-state waveforms in a numerical, point source model in a half-space environment. It was found that TR can effectively localize a simple source broadcasting a continuous wave, depending on the angular spacing. Furthermore, the angular spacing and the aperture of the TRM are the most important parameters when creating a setup of receivers for imaging a source. That work optimizes a TRM when the source's location is known within a region of certainty.

Ibrahim El Baba et al [10] have mentioned an original way of using TR in electromagnetics is detailed (principally concerning the ability of the technique to focus a signal in time and space). First, the theoretical principles of TR were described and illustrated using Finite Difference in Time Domain computing of Maxwell's equations. Then, the aim was to accurately describe the influence of various parameters (from FDTD numerical experiments) above focusing. Thus, particular interests rely on the numbers and locations of TR sensors, directivity of source, presence of scatterers, and impact of MSRC. Finally, considering characteristics of EMC applications in MSRC, a closer look has been set to the advantages of TR numerical tools for innovating studies in reverberation chambers.

2.4 Finite Difference Time Domain Method

The Finite Difference Time Domain (FDTD) method is an application of the finite difference method, commonly used in solving differential equations, to solve Maxwell's equations. In FDTD, space is divided into small portions called cells. On the surfaces of each cell, there are assigned points. Each point in the cell is required to satisfy Maxwell's equations. In this way, electromagnetic waves are simulated to propagate in a numerical space, almost as they do in real physical world. FDTD is one of the commonly used methods to analyse electromagnetic phenomena at radio and microwave frequencies. Computer programs written in MATLAB can display electromagnetic phenomena in movies.

Electromagnetic is governed by the time-dependent Maxwell's curl equations, which in free space are

$$\frac{\partial E}{\partial t} = \frac{1}{\varepsilon_0} \nabla \times H \qquad (2.1\ a)$$

$$\frac{\partial H}{\partial t} = -\frac{1}{\mu_0} \nabla \times E . \qquad (2.1\ b)$$

E and H are vectors in three dimensions, but if we consider only one dimension

$$\frac{\partial E_x}{\partial t} = -\frac{1}{\varepsilon_0} \frac{\partial H_y}{\partial z} \qquad (2.2\ a)$$

$$\frac{\partial H_y}{\partial t} = -\frac{1}{\mu_0} \frac{\partial E_x}{\partial z} . \qquad (2.2\ b)$$

To put these equations in a computer, we approximate the derivatives with the "finite-difference" approximations:

$$\frac{E_x^{n+1/2}(k) - E_x^{n-1/2}(k)}{\Delta t} = -\frac{1}{\varepsilon_0} \frac{H_y^n(k+1/2) - H_y^n(k-1/2)}{\Delta x} \qquad (2.3\ a)$$

$$\frac{H_y^{n+1}(k+1/2) - H_y^n(k+1/2)}{\Delta t} = -\frac{1}{\mu_0} \frac{E_x^{n+1/2}(k+1) - E_x^{n+1/2}(k)}{\Delta x} . \qquad (2.3\ b)$$

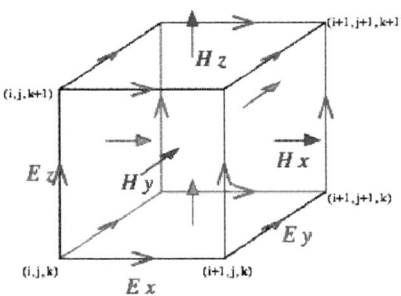

Figure.2.1. Illustration of a standard Cartesian Yee cell used for FDTD, about which electric and magnetic field vector components are distributed. Visualized as a cubic voxel, the electric field components form the edges of the cube, and the magnetic field components form the normals to the faces of the cube. A three-

12

dimensional space lattice consists of a multiplicity of such Yee cells. An electromagnetic wave interaction structure is mapped into the space lattice by assigning appropriate values of permittivity to each electric field component, and permeability to each magnetic field component.

In these two equations, time is specified by the superscripts, i. e., "n" actually means a time $t = \Delta t \cdot n$, and "k" actually means the distance $z = \Delta x \cdot k$. (It might seem more sensible to use Δz as the incremental step, since in this case we are going in the z direction. However, Δx is so commonly used for a spatial increment that we will use Δx.). We rearrange the above equations to :

$$E_x^{n+1/2}(k) = E_x^{n-1/2}(k) - \frac{\Delta t}{\varepsilon_0 \cdot \Delta x} \left[H_y^n(k+1/2) - H_y^n(k-1/2) \right] \qquad (2.4\ a)$$

$$H_y^{n+1}(k+1/2) = H_y^n(k+1/2) - \frac{\Delta t}{\mu_0 \cdot \Delta x} \left[E_x^{n+1/2}(k+1) - E_x^{n+1/2}(k) \right]. \qquad (2.4\ b)$$

Notice that the calculations are interleaved in both space and time. In Eq. (2.4 a), for example, the new value of E_x is calculated from the previous value of E_x and the most recent values of H_y. This is the fundamental paradigm of the finite-difference time-domain (FDTD) method Figure.2.2).

Eq. (2.4 a) and (2.4 b) look very similar. However, ε_0 and μ_0 differ by several orders of magnitude:

$$\varepsilon_0 = 8.85 \times 10^8 \quad F/m, \quad \mu_0 = 4\pi \times 10^{-7} \quad H/m.$$

Therefore, E_x and H_y will differ by several orders of magnitude. This is circumvented by making the following change of variables:

$$\tilde{E} = \sqrt{\frac{\varepsilon_0}{\mu_0}} E. \qquad (2.5)$$

Substituting this into Eq. (2.4a) and (2.4b) gives

$$\tilde{E}_x^{n+1/2}(k) = \tilde{E}_x^{n-1/2}(k) - \frac{1}{\sqrt{\varepsilon_0 \mu_0}} \frac{\Delta t}{\Delta x} \left[H_y^n(k+1/2) - H_y^n(k-1/2) \right] \qquad (2.6a)$$

$$H_y^{n+1}(k+1/2) = H_y^n(k+1/2) - \frac{1}{\sqrt{\varepsilon_0 \mu_0}} \frac{\Delta t}{\Delta x} \left[\tilde{E}_x^{n+1/2}(k+1) - \tilde{E}_x^{n+1/2}(k) \right] \qquad (2.6b)$$

13

Now both $\tilde{E}$ and H will have the same order of magnitude. We will call this "normalized" units. Physicist call this Gaussian units. Note that

$$\left[\sqrt{\frac{\varepsilon_0}{\mu_0}}\right] = \left[\frac{F/m}{H/m}\right]^{1/2} = \left[\frac{C/V}{Wb/A}\right]^{1/2} \qquad \text{and}$$

$$= \left[\frac{C/V}{V/s-A}\right]^{1/2} = \left[\frac{A}{V}\right] = \left[\frac{1}{ohm}\right]$$

$$\sqrt{\frac{\mu_0}{\varepsilon_0}} = \sqrt{\frac{4\pi \times 10^{-7}}{8.85 \times 10^{-12}}} = \sqrt{1.42 \times 10^5} = 377\,\Omega$$

This quantity is called the "impedance of free space."

Once the cell size Δx is chosen, then the time step Δt is determined by

$$\Delta t = \frac{\Delta x}{2 \cdot c_0} \tag{2.7}$$

where c_0 is the speed of light in free space. Therefore,

$$\frac{1}{\sqrt{\varepsilon_0 \mu_0}} \frac{\Delta t}{\Delta x} = c_0 \cdot \frac{\Delta x / 2 \cdot c_0}{\Delta x} = \frac{1}{2} \tag{2.8}$$

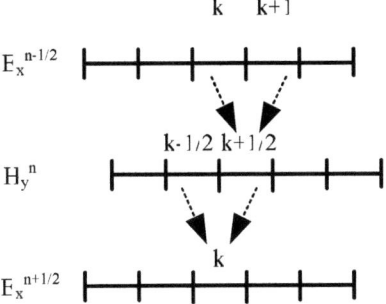

Figure.2.2. A Diagram of the Calculation of E and H Fields in FDTD

2.5 Proposed Model

This research is acoustics simulation to test the ideas behind time reversal methods for communications. The program is complete graphics based and allows users to control many different aspects of the program. The size of the simulation, the placement of the receiver and array can be defined.

14

Several preset environments can be chosen, or the user can input their own environment by hand. The user can test three types of digital modulation, ASK, PSK, and FSK. The full signal constellation is transmitted to the array. The array reversed the signal in time and stores it as the new prototype signal constellation. The user can then transmit the prototype signals back to the x through the same environment. Throughout the process, the electric field intensity is plotted on a log scale. The entire process can be recorded and then converted to an AVI file. The acoustics simulation is performed using the FDTD algorithm with PML boundary conditions.

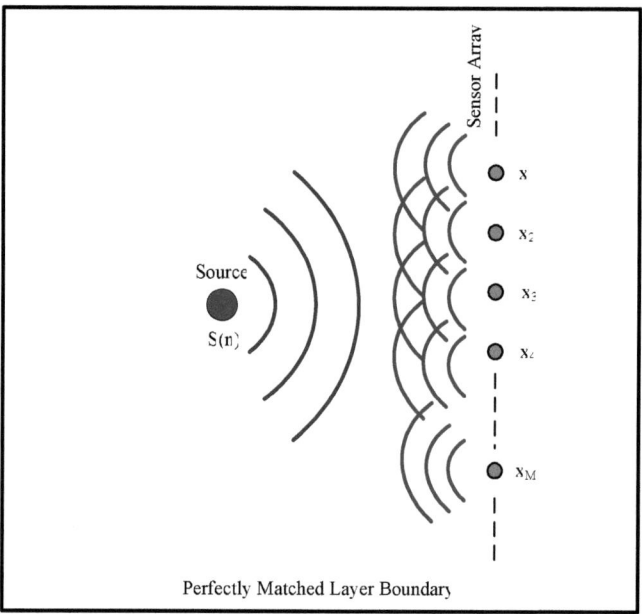

Figure.2.3. Proposed Simulation Model

2.6 Room Impulse Responses Model Development

Real room acoustic impulse responses (AIRs) modelled by infinite impulse response (IIR) filters require high model orders. Many problems involving the estimation of AIRs reduce to high dimensional optimisation problems. The transfer function due to the acoustics of a room generally does not change considerably with time, but do vary with the spatial locations of the sound source and observer. Assuming both are spatially stationary, a linear time-invariant (LTI) model is appropriate. The all-pole model can parsimoniously approximate rational transfer functions, and typical all-pole model orders required for approximating room transfer functions (RTFs) are in the range $50 \leq P \leq 500$ - around a factor of 40 lower than all-zero model orders.

A room acoustic impulse response (AIR), rir(t), may be modelled by a LTI all-pole filter of order P, as given by:

$$rir(t) = -\sum_{p \in P} a(p)rir(t-p) + \delta(t), \ t \in Z \qquad (2.9)$$

where $a = \{a(p), p \in P \triangleq \{1,...,P\}\}$ are the model parameters, P is the number of poles, and $\delta(t)$ is the Kronecker delta.

In many applications, such as single channel blind dereverberation, an estimate of the AIR is required and, in general, this reduces to a high-dimensional optimisation problem. This is difficult to solve because attempts to model the entire acoustic spectrum by a single W filter leads to a large computational load, as well as numerical problems resulting from the size of the parameter space [11].

Chapter 3

Implementation and Simulation Results

The system block diagram is illustrated in Figure 3.1. There are six main processes to develop the proposed system. In the simulation setup stage, the size of boundary, the location of source or transmitter, the location of transducer array and choice of medium have to specify. After finishing the simulation setup stage, the modulation schemes such as amplitude shift keying, frequency shift keying and phase shift keying have to be selected. And then the signal is transmitted from the source first. This signal can propagate to the transducer array through the specified medium. That signal must be time reversed by transducer array. And then the time reversed signals have to be re-transmitted to the source. The location of source can be found by receiving the re-transmitted signals.

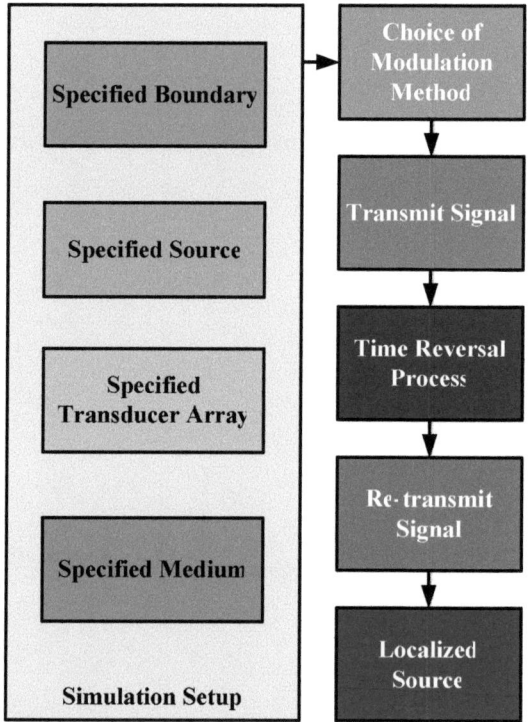

Figure 3.1. Proposed Block Diagram

3.1 Software Development

We have implemented to develop the time-reversal focusing was designed for human computer interface applications by exploiting the ultra-wideband (UWB) acoustics waves based on the structure of Figure.3.2.

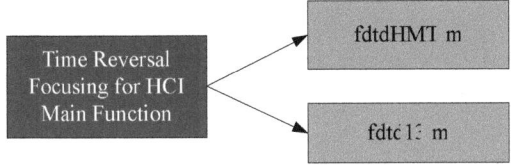

Figure.3.2. Linking Function of MATLAB GUI Implementation

3.2 Overall System Flowchart

We have to define some variables to set up for time reversal focusing simulation environments.

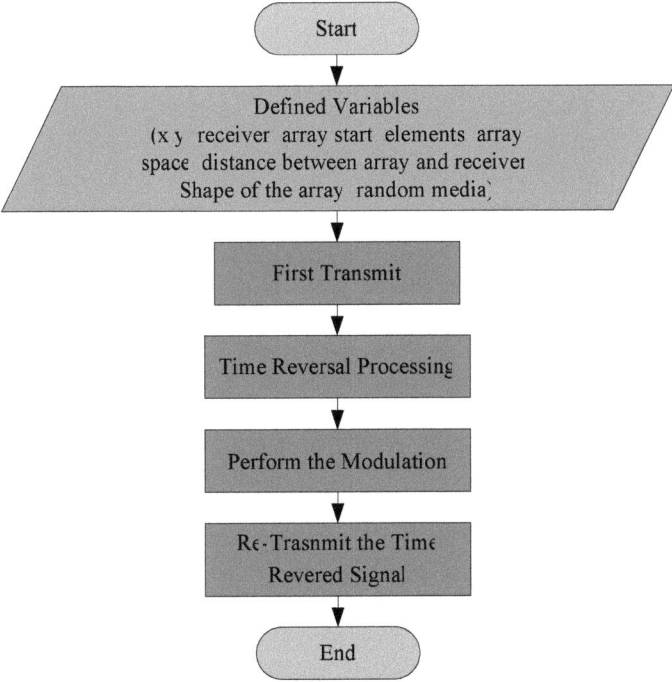

Figure.3.3. System Flowchart

The x, y, array start, elements, array space, distance between array and receiver, shape of the array and random media for proposed scenario. In a time reversal experiment, the medium covering the targets is illuminated by an array of transmitters. The signal propagates through the medium, interacts with the targets and the return echoes are recorded by an array of receivers called time reversal array or time reversal mirror (in the generalized case the transmitter and receiver arrays can be separated). These recorded signals are reversed in time and sent back to the probing medium where they will experience all the events that scattered signals faced and finally focus on the position of the targets. The overall flowchart is shown in Figure 3.3.

3.2.1. Specified Boundary

The input parameters for boundary specification have been defined based on the following factors:

- imax: size of grid in x-direction
- jmax: size of grid in y-direction
- del: size of cell in meters
- signal: matrix with row vectors of input signal
- space: the number of cells between transmitters
- start: the cell at which the first transmitter is located
- array: the number of elements in the receiving array
- rstart: the x-coordinate of the first receiver
- rspace: the spacing between receivers
- depth: the distance between the receiving array and the transmitting array
- offset: the distance from zero in the x-direction where the simulation will start
- media: value between 1 and 5 that determines the type of environment in which the experiment takes place
- ER: when media =5, a user defined environment is provided

The output parameters for boundary specification have been defined based on the following factors:

- E: Outputs the final electric field
- V: Outputs the received signal at certain points

3.2.2. Specified Source

In source localization algorithm, the developed simulation with the help of MATLAB can be defined based on the x and y grid and the following instructions to create the specified grid size for source. The receiver can be placed by using the location algorithm based on defining source location in simulation environments.

handles.x= str2double(get(h,'string'));

handles.y = str2double(get(h,'string'));

handles.receiver=str2double(get(h,'string'));

3.2.3. Specified Transducer Array

In specified transducer array, the developed simulation with the help of MATLAB can be defined based on the array start, elements, array space, offset and the following instructions to create the specified transducer array. The transducer array can be placed by using the location algorithm based on defining transducer array in simulation environments.

handles.array_start = str2double(get(h,'string'));

handles.elements = str2double(get(h,'string'));

handles.arrayspace = str2double(get(h,'string'));

handles.offset = str2double(get(h,'string'));

3.2.4. Specified Medium

There are five mediums to select the specific application of time reversal mirror based source localization.

- Waveguide
- Line of sight

- No line of sight
- Free Space
- User Defined

3.2.5. Choice of Modulation Method

There are four modulation schemes to analyze the source localization of time reversal signal processing for time reversal mirror implementation.

- No Modulation
- Amplitude Shift Keying
- Phase Shift Keying
- Frequency Shift Keying

3.2.6. Transmit Signal

After defining the source and transducer array location, the transmit signal from the transmitter according to the predefined process of fdtd13.m. The signal created by the predefined process of fdtd13.m is analyzed by using finite difference time domain algorithm for proposed source localization algorithm.

3.2.7. Time Reversal Process

In time reversal experiment, a short pulse transmitted by a source through a high scattering medium is received by an array, then time reversed, energy normalized, and retransmitted through the same medium. If the scattering channel is reciprocal and rich in multipath, the time reversed signals travel backwards through the background medium and undergo similar changes (such as multiple scattering, reflections, and refraction) that they underwent in the forward direction step and finally refocuses on the original source.

3.2.8. Re-transmit Signal

After defining the source and transducer array location, the re-transmit signal from the transducer array can be transmitted back according to the predefined process of fdtd13.m. The re-transmit signal created by the predefined process of fdtd13.m is analyzed by using finite difference time domain algorithm for proposed source localization algorithm.

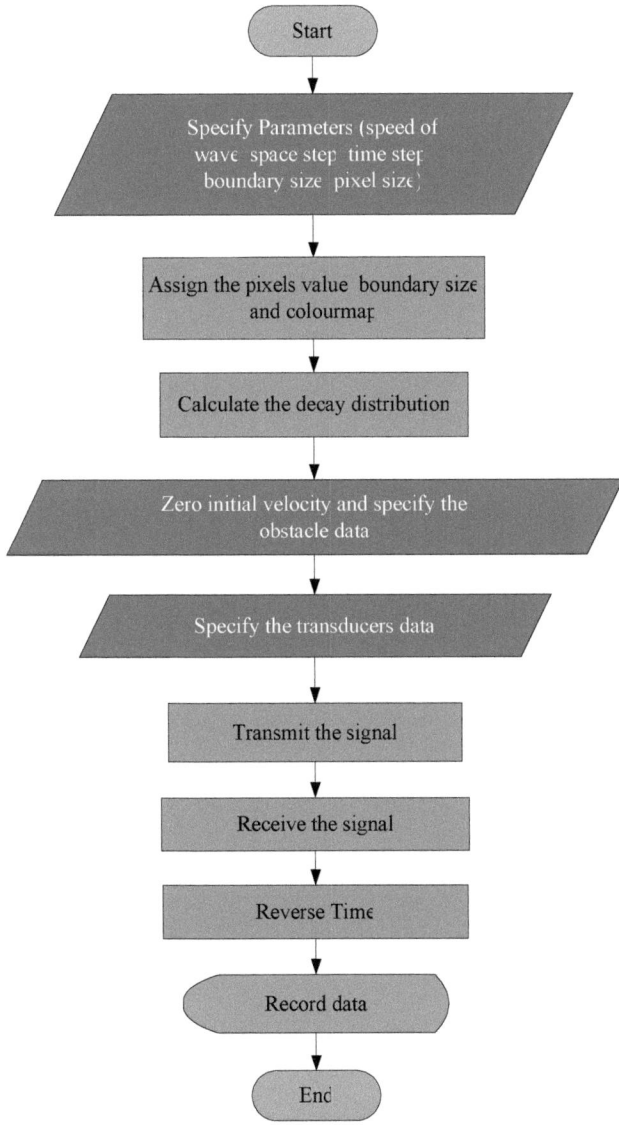

Figure 3.4. Flowchart of Time Reverse Process

3.2.9. Localized Source

When the re-transmit signal from transducer array can be localized the location of source, the x and y coordinate for source are expressed in command window for user. The exact position of source can be detected by using the proposed localization algorithm.

3.3. Development of Room Impulse Response

The flowchart of room impulse response model is illustrated in Figure 3.5. The various parameters such as room information and the impulse response are declared to get the room impulse model design. The index of the sequence to obtain the appropriate room model has to be evaluated. The vectors for acoustics signal are converted to 3D matrices to attain the specified impulse response. According to the parameters of room model and evaluation of acoustics signal, the room impulse response is displayed.

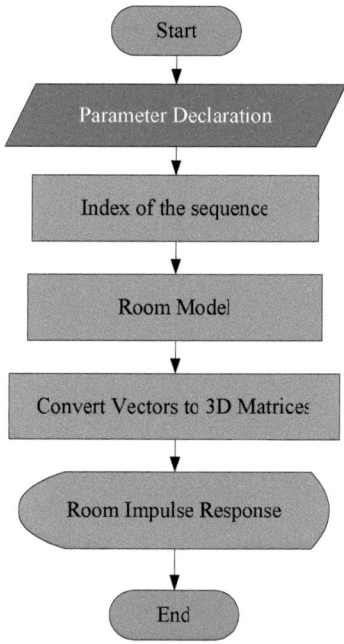

Figure 3.5. Flowchart of Room Impulse Response Model

3.4. Flowchart of FDTD Scheme

Figure 3.6 shows the flowchart of finite different time domain. In this flowchart, there are several input parameters to declare and operate the simulation of FDTD scheme. They are domain length, spatial samples in domain, iteration, source frequency, spatial step, magnetic time step, permittivity, permeability, x coordinate of spatial samples, WKB approximation. The scale factors for electric and magnetic fields can be calculated by utilizing the declared input parameter. After getting the calculated result from the above calculation, the input to the profile for decision of separate processes can be acquired.

If there are profiles, the profile numbers can be checked from 1 to 6 to evaluate the difference processes when the condition is true. If profile 1 is factual, the dielectric window can be displayed. If profile 2 is accurate, the dielectric window with smooth transition can be shown. If profile 3 is right, the dielectric discontinuity can be exhibited. If profile 4 is spot on, the dielectric discontinuity with ¼ wave matching layer can be presented. If profile 5 is correct, the conducting half space can be demonstrated. If profile 6 is proper, the sinusoidal dielectic can be put on show. And then the electric field and the magnetic field can be calculated from the above six profile statement. And then the permittivity, permeability, conductivity profiles are plotted. If the value of field is zero, the absorbing boundary conditions for left propagating waves and right propagating waves to update electric field and magnetic field will be evaluated. And then the electric and magnetic field can be monitored with the help of MATLAB spectrum for localization. If there are no profiles, the program will be terminated.

3.5. Analysis of Source Localization for Proposed Research Work

Figure.3.7 illustrates the flowchart of time reversal mirror for source localization. In this flowchart, there are several input parameters to declare and analyze the source localization for proposed research work. They are X direction, Y direction, receiver in x, location of first array elements, elements in the array, spacing between array elements, distance between array and receiver, offset from x = 0 and receiver, shape of the array, environment.

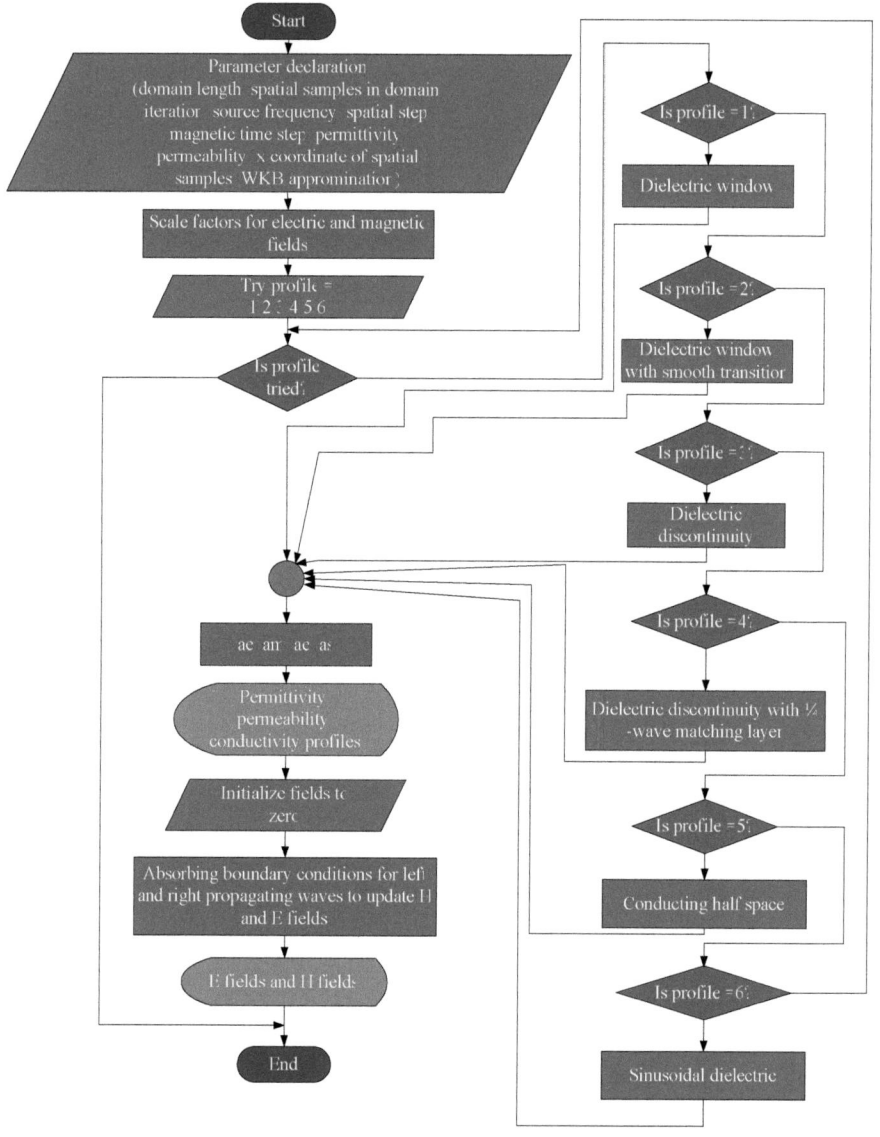

Figure.3.6 .Flowchart of Finite Different Time Domain

There are five predefined process for this analysis. They are first transmission, time reversal, perform the modulation, re-transmit the time reversed signal, and show the simulation setup. After completing the predefined processes, the program will be finished and the analysis of source localization for proposed research work will be accomplished.

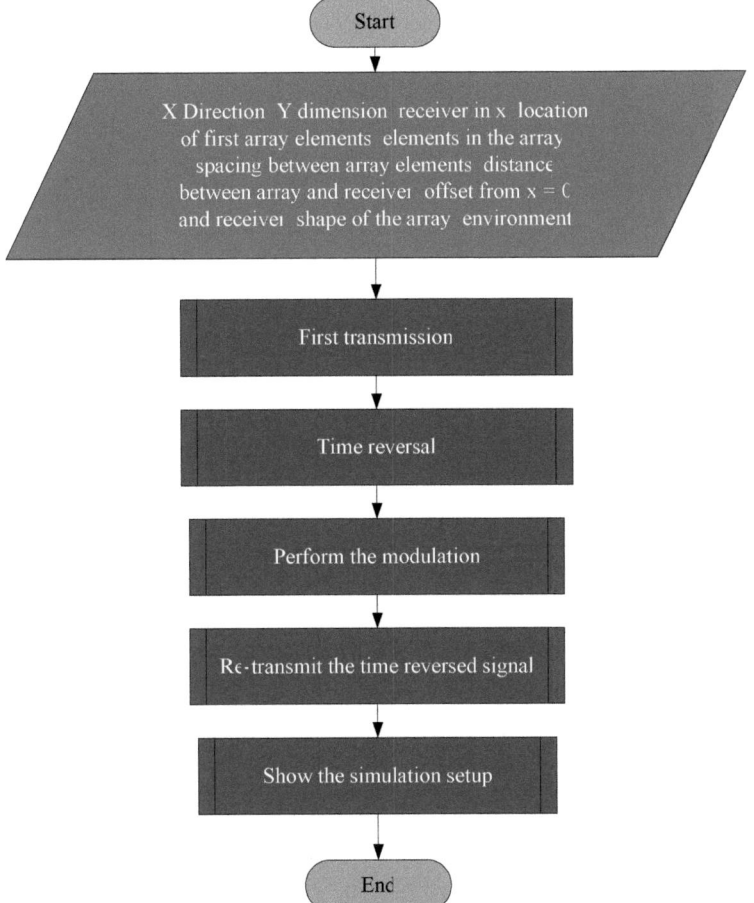

Figure.3.7 . Flowchart of Time Reversal Mirror for Source Localization

Figure.3.8 demonstrates flowchart of first transmission. The first transmission predefined process is initialized. In this flowchart, there are several input parameters to declare and first transmission for proposed research work. They are X, Y, depth, offset, array start, array space, elements, receiver, media, fc, c, , del, time, step, fs. The signals to transmit from transmitter are plotted after getting the input parameter. The original signal from the receiver can be transmitted and the predefined process of fdtd13 can be activated based on the transmitted signal. The result data from fdtd13 can be returned to specified called function.

Figure.3.9 shows flowchart of fdtd13. The fdtd13 predefined process is initialized. In this flowchart, there are several input parameters to declare and fdtd13 for proposed research work. They are Imax, jmax, del, signal, space, start, array, rstart, rspace, depth, offset, media, ER, Constitutive parameters (eo, uo, c), time step, input of several signals at a time, PML :number of layers. The total grid size including PML layer can be computed by using the above parameters. And the forward propagation and reverse propagation are also computed.

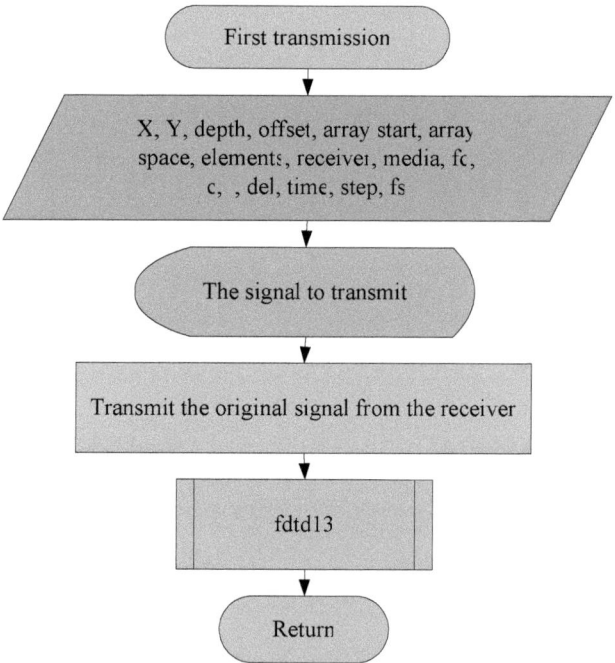

Figure.3.8. Flowchart of First Transmission

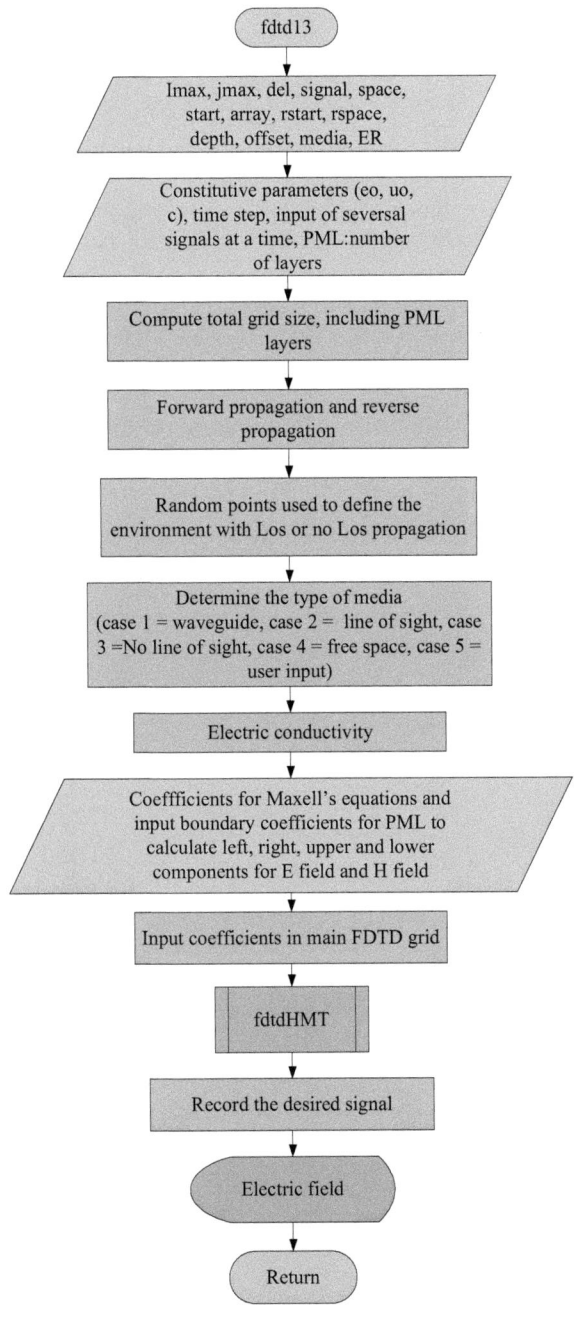

Figure.3.9. Flowchart of fdtd13

The random points used to define the environment with LOS or no LOS propagation is evaluated. The type of media (case 1 = waveguide, case 2 = line of sight, case 3 =No line of sight, case 4 = free space, case 5 = user input) are determined. After determining the environments, the electric conductivity can be calculated.

The coefficients for Maxell's equations and input boundary coefficients for PML to calculate left, right, upper and lower components for E field and H field are defined. After defining the coefficient, the input coefficients in main FDTD grid can be evaluated. And then the predefined process fdtdHMT can be called from fdtd13 function. The desired signal from fdtdHMT can be recorded and the electric field spectrum can be displayed.

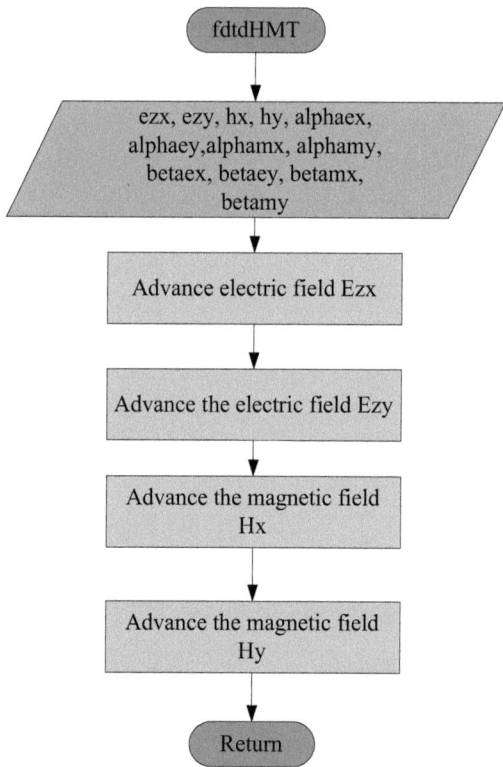

Figure.3.10. Flowchart of fdtdHMT

Figure.3.10 demonstrates the flowchart of fdtdHMT. The input parameters are ezx, ezy, hx, hy, alphaex, alphaey,alphamx, alphamy, betaex, betaey, betamx, betamy. After declaring the input parameters, the advanced electric field Ezx, Ezy, the advanced magnetic field Hx, Hy, are calculated. After getting the calculated parameter of the advanced electric field Ezx, Ezy, the advanced magnetic field Hx, Hy, the program will be returned to called function.

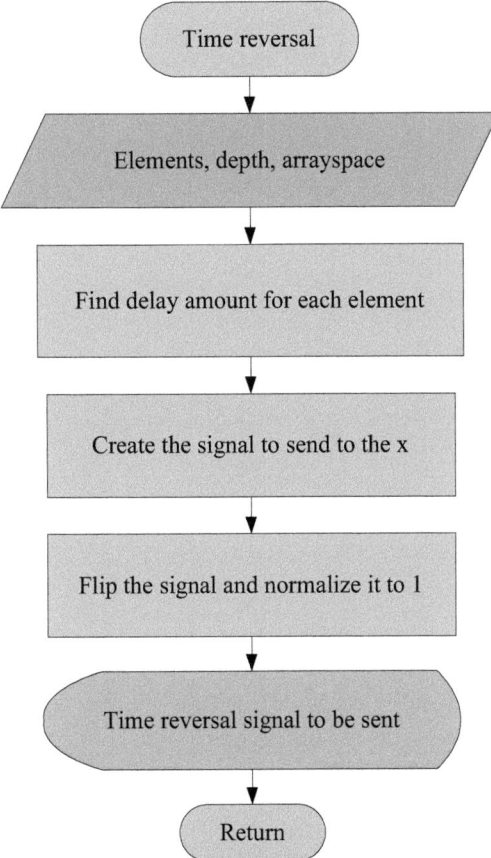

Figure.3.11. Flowchart of Time Reversal

Figure.3.11 shows the flowchart of Time Reversal. The input parameters are Elements, depth, arrayspace. The delay amount for each element is found based on the input parameters. The signal to send to the x direction can be created and the normalized signal to 1 is flipped. The time reversed signal to be sent from transducer array will be displayed and the signal values are returned to the called function.

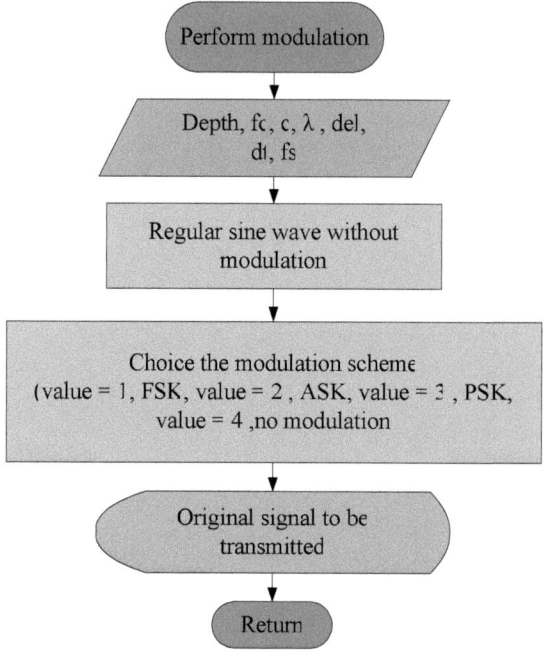

Figure.3.12 . Flowchart of Perform the Modulation

Figure.3.12 illustrates the flowchart of Perform the Modulation. The input parameters are Depth, fc, c, λ , del, dt, fs. The regular sine wave without modulation is developed. The modulation scheme (value = 1, FSK, value = 2 , ASK, value = 3 , PSK, value = 4 ,no modulation is selected. The original signal to be transmitted from source will be displayed and returned to the called function.

Figure.3.13 mentions the flowchart of Retransmit the Time Reversed Signal. The input parameters are X, Y, del, send, array start, array space, depth, offset, media. The predefined process of Transmit (fdtd13) can be called and the received signals from the array are displayed.

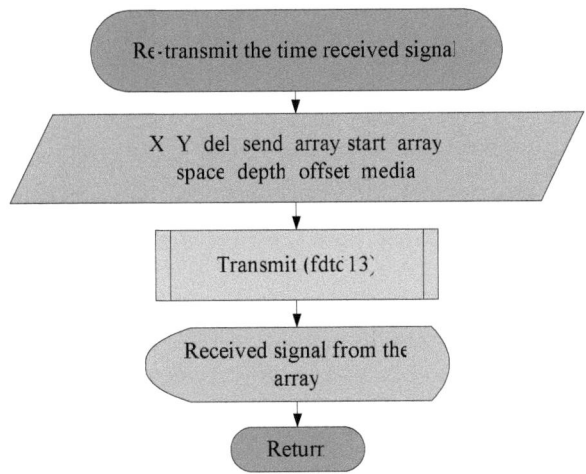

Figure.3.13. Flowchart of Retransmit the Time Reversed Signal

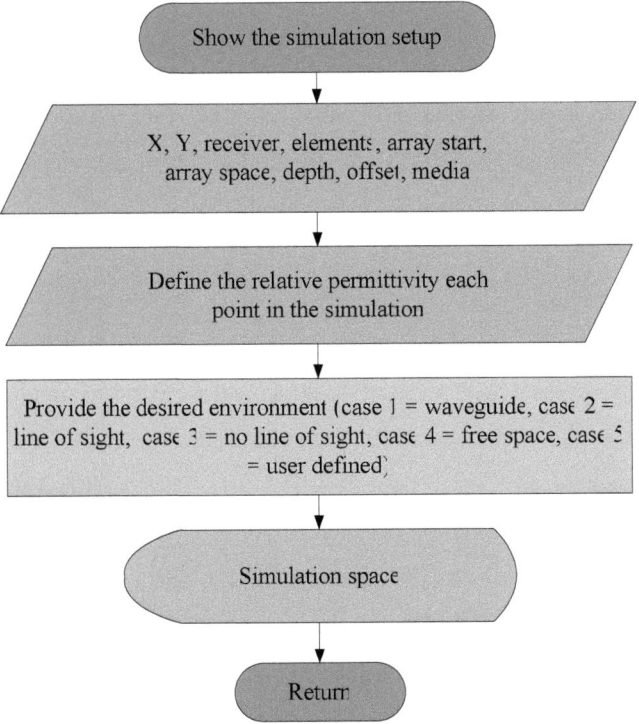

Figure.3.14. Flowchart of Show the Simulation Setup

Figure.3.14 shows the flowchart of Show the Simulation Setup. The input parameters are X, Y, receiver, elements, array start, array space, depth, offset, media. The relative permittivity each point in the simulation can be defined. The desired environment (case 1 = waveguide, case 2 = line of sight, case 3 = no line of sight, case 4 = free space, case 5 = user defined) are provided. The simulation space will be displayed and the result can be returned to the called function.

3.6. Analysis of Bit Error Rate

The bit error rate analysis for various medium for simulation approach on time reversal mirror is evaluated as the following equation:

$$\text{BER } (H,P) = \sum_{l=1}^{M^B} \frac{P_r(x_l)}{B \log_2(M)} \sum_{k=1, \, k \neq l}^{M^B} P_{r_{HP}(\check{x}_k| \, x_l)} \, d(\check{x}_k, x_l) \tag{3.1}$$

where

$\check{x}_k$ is the estimated symbol sequence, x_l is the transmitted symbol sequence, $P_{rHP}(\check{x}_k, x_l)$ is the probability of deciding as $\check{x}_k$ when x_l is transmitted, BER(H,P) is the BER for a Room Model H, and Medium P, and $d(\check{x}_k, x_l)$ is the number of bit errors incurred in deciding as x_k when x_l has been transmitted.

3.7 Simulation Results

The experimental setup for boundary condition is specified with the range of field setup. The x, y and receiver location from field setup in the simulation window are 100m, 100m, and 50m respectively. Figure.3.15 illustrates the experimental setup for eight elements in free space condition. The array start at 18 and the number of elements are 8 for this analysis. The number of cells between transmitters called space may be located by 10m. The distance between transmitter and array is about 90m and the offset value is also 10m for the location of setup.

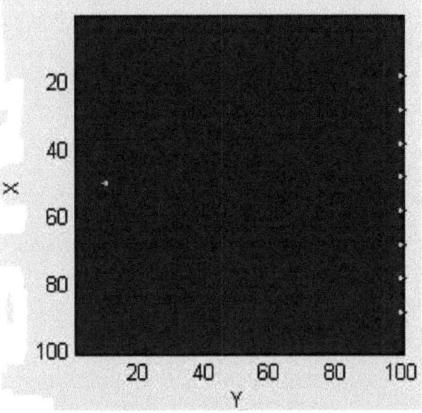

Figure.3.15. Experimental Setup for Eight Elements in Free Space Model

After defining the specified parameters for experimental setup, there are many modulation schemes for ping stage. The present simulation test is chosen Amplitude Shift Keying (ASK) modulation. Figure.3.16 shows the screenshot result for ASK modulation in free space condition.

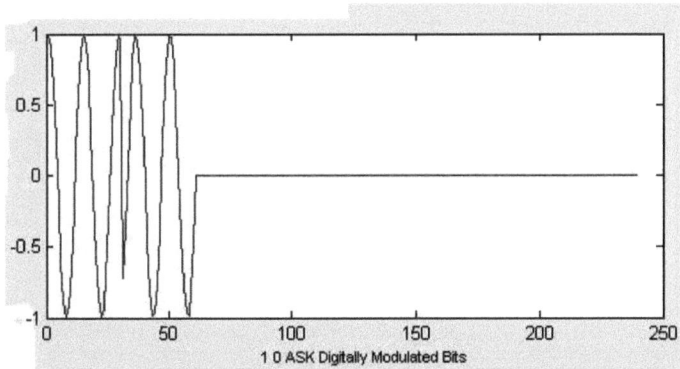

Figure.3.16. Screenshot Result for ASK Modulation in Free Space Model

The transmitter starts to transmit the acoustics signal to the array when the ping button is pressed. Figure.3.17 mentions the screenshot results for ping stage in free space condition. The red colour shows the high intensity level and the gray-blue colour means the low intensity level of signal energy.

34

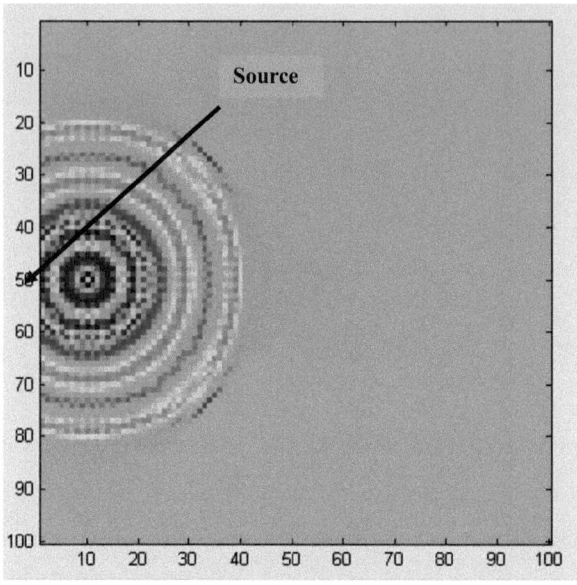

Figure.3.17. Screenshot Results for Ping Stage in Free Space Model

The acoustics wave is scattered to the array of transducers and the waves are propagated to reach the perfect boundary. Figure.3.18 illustrates the screenshot result for received ping at the array elements.

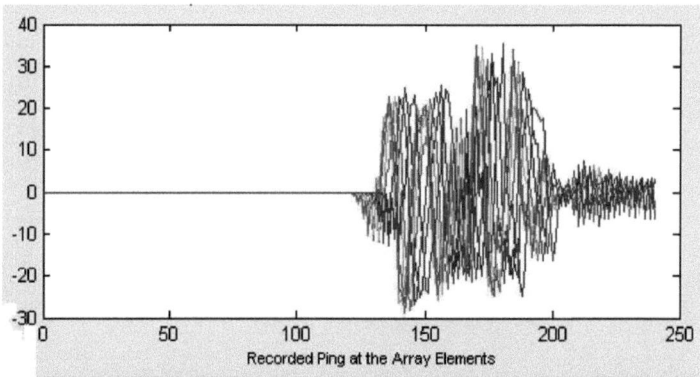

Recorded Ping at the Array Elements

Figure.3.18. Screenshot Result for Received Ping at the Array Elements

After receiving the propagated signal to the array of transducers, the time reversed signal can be retransmitted from the array of transducers to the transmitter by utilizing the time reversed button from the panel. Figure.3.19 shows the screenshot result for time reversed signal to be sent.

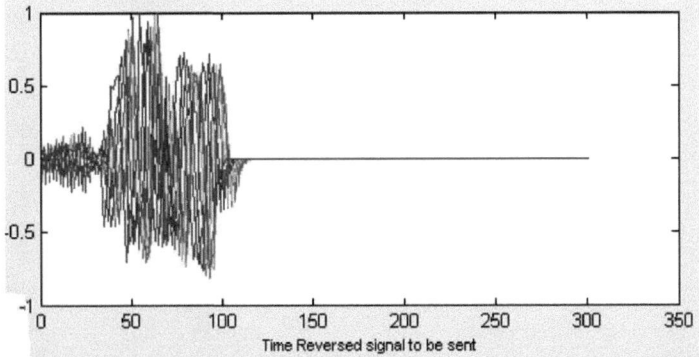

Figure.3.19. Screenshot Result for Time Reversed Signal to be Sent

The propagation of the acoustics wave from the array of transducers to the transmitter is illustrated in Figure.3.20. The waves may be scattered to the boundary and reached to the transmitter in free space stage.

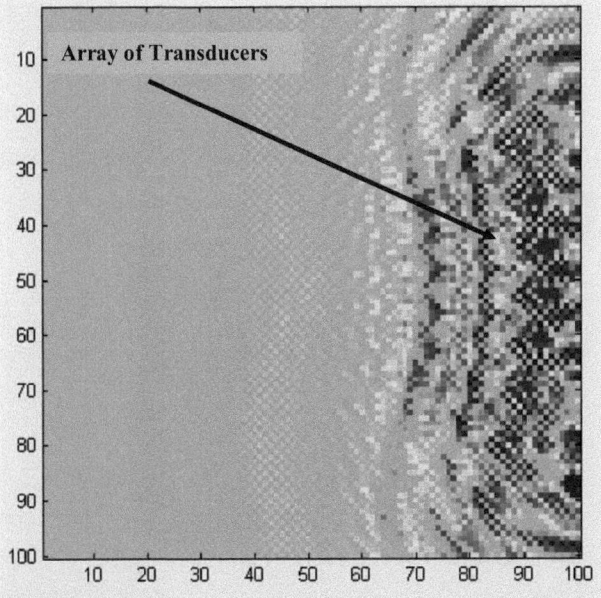

Figure.3.20. Screenshot Result for Retransmit Stage from Array in Free Space Model

Figure.3.21 mentions the screenshot result for received signal from the array for the free space analysis.

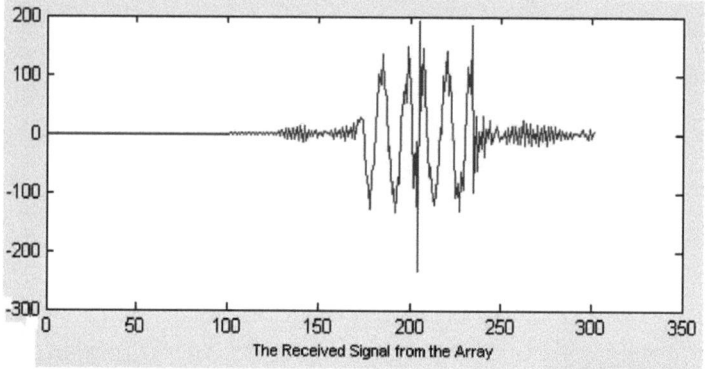

The Received Signal from the Array

Figure.3.21. Screenshot Result for Received Signal from the Array

The investigational arrangement for boundary condition is specified with the range of field setup for waveguide situation. The x, y and receiver location from field setup for waveguide stage in the simulation window are 100m, 100m, and 50m respectively. Figure.3.22 demonstrates the experimental setup for eight elements in waveguide condition. The array start at 18 and the number of elements are 8 for this waveguide condition. The number of cells between transmitters called space may be positioned by 10m. The distance between transmitter and array is about 90m and the offset value is also 10 for the location of waveguide condition.

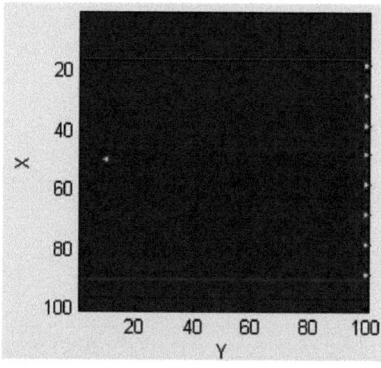

Figure.3.22 Experimental Setup for Eight Elements in Waveguide Model

After defining the specified parameters for experimental setup for waveguide condition, there are various modulation schemes for ping stage. The second simulation test is chosen Amplitude Shift Keying (ASK) modulation. The transmitter starts to transmit the acoustics signal to the array when the ping button is pressed. Figure.3.23 illustrates the screenshot results for ping stage in waveguide condition. The acoustics wave is scattered to the array of transducers and the waves are spread to reach the perfect boundary.

Figure.3.24 points up the screenshot result for received ping at the array elements in waveguide condition. After receiving the propagated signal to the array of transducers, the time reversed signal can be retransmitted from the array of transducers to the transmitter by exploiting the time reversed button from the panel. Figure.3.25 shows the screenshot result for time reversed signal to be sent for waveguide condition.

The propagation of the acoustics wave from the array of transducers to the transmitter is demonstrated in Figure.3.26. The waves may be scattered to the boundary and achieved to the transmitter in waveguide condition.

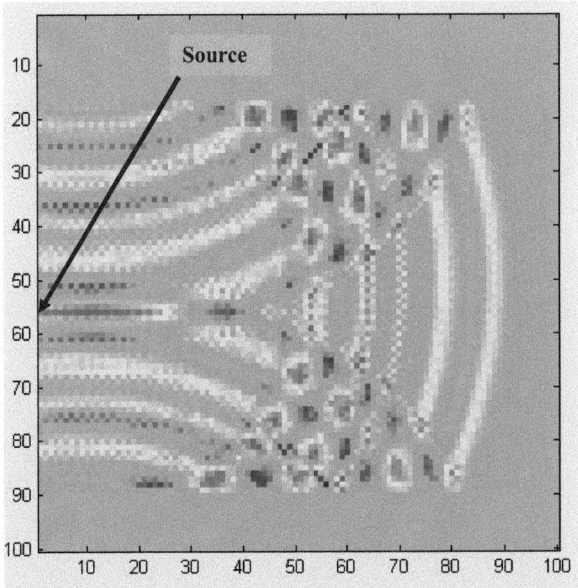

Figure.3.23. Screenshot Result for Ping Stage in Waveguide Model

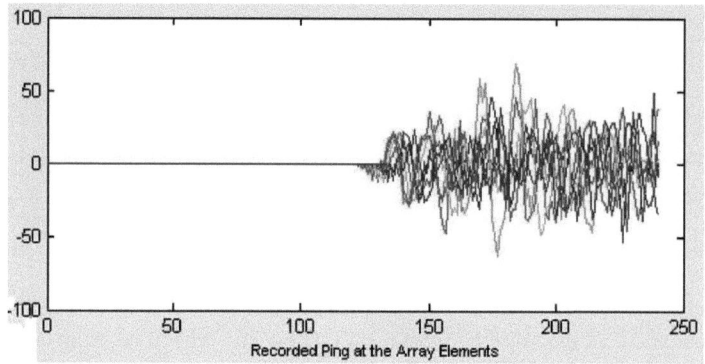

Figure.3.24. Screenshot Result for Received Ping at the Array Elements

Figure.3.27 shows the screenshot result for received signal from the array for the waveguide condition. By comparing with the waveguide condition, the span of received signal is wider than the first simulation result.

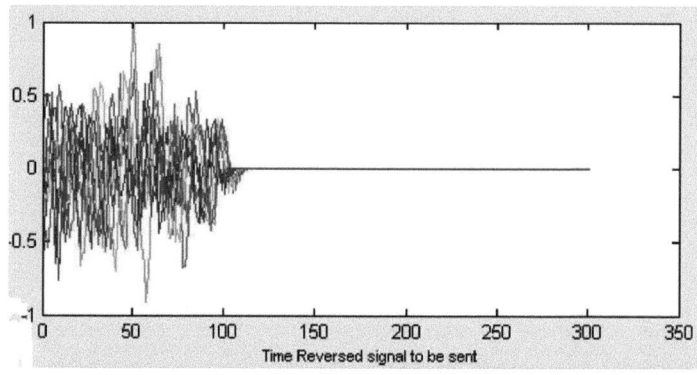

Figure.3.25. Screenshot Result for Time Reversed Signal to be Sent

The last simulation test for time reversal focusing is based on the user defined stage. The user may determine which type of environment to run the simulation under the selection of needs. The six objects may be positioned between the transmitter and the array of transducers as shown in Figure.3.28.

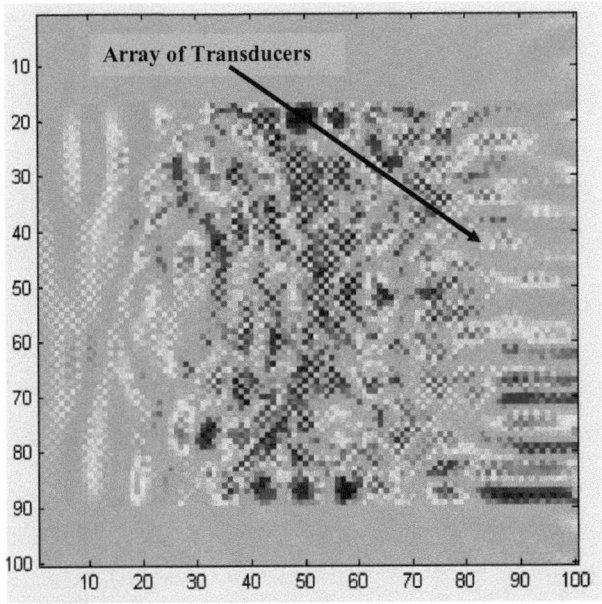

Figure.3.26. Screenshot Result for Retransmit Stage from Array in Waveguide Model

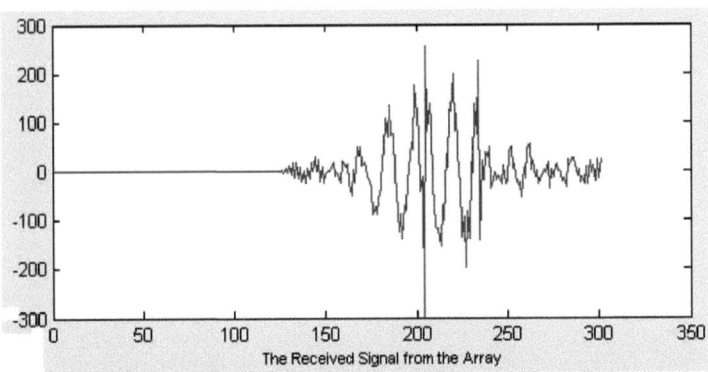

The Received Signal from the Array

Figure.3.27. Screenshot Result for Received Signal from the Array

This is the experimental setup for eight elements in user defined stage of specified six objects between the transmitter and the array of transducers.

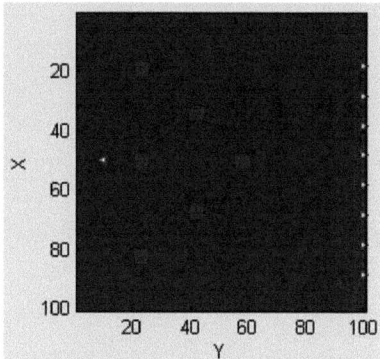

Figure.3.28. Experimental Setup for Eight Elements of TRM array in User Defined Model. The Red Boxes Indicate User Defined Location for Finding the Intensity.

After declaring the specified parameters for experimental setup for user defined condition, there are different modulation schemes for ping stage. The last simulation test is preferred Amplitude Shift Keying (ASK) modulation. The transmitter starts to transmit the acoustics signal to the array when the ping button is pressed.

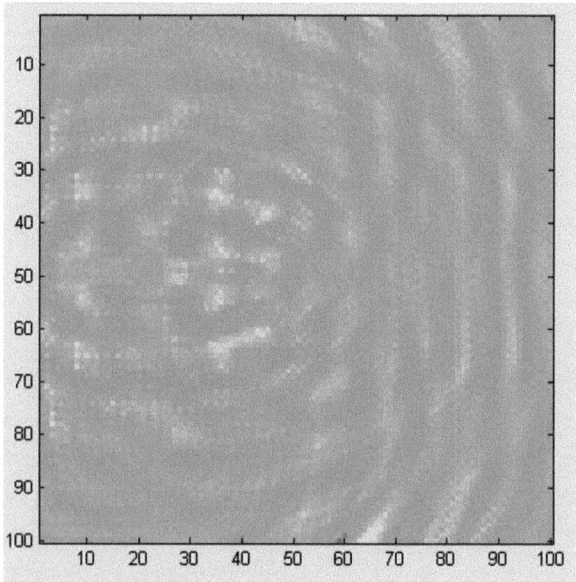

Figure.3.29. Screenshot Result for ASK Modulation in User Defined Model

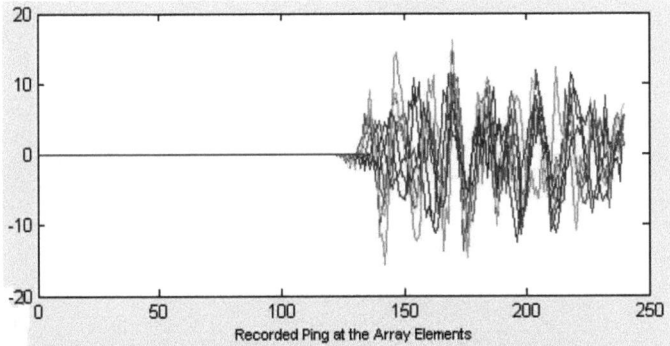

Figure.3.30. Screenshot Result for Received Ping of the Array Elements

Figure.3.29 illustrates the screenshot results for ping stage in user defined condition. The acoustics wave is sprinkled to the array of transducers and the waves are stretch to get to the perfect boundary. Figure.3.30 emphasizes the screenshot result for received ping at the array elements in user defined condition.

After receiving the proliferated signal to the array of transducers, the time reversed signal can be retransmitted from the array of transducers to the transmitter by using the time reversed button from the panel. Figure.3.31 shows the screenshot result for time reversed signal to be sent for user defined condition.

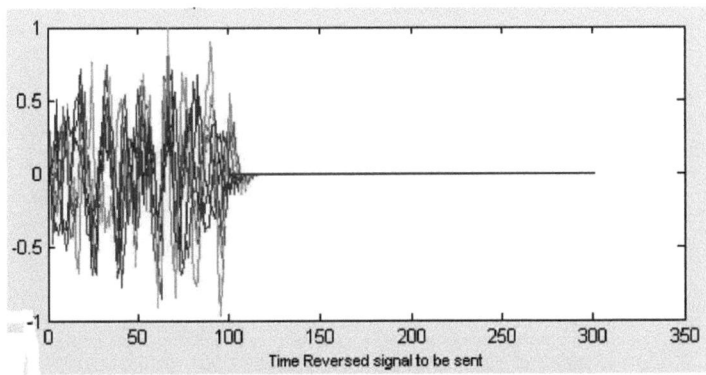

Figure.3.31. Screenshot Result for Time Reversed Signal to be Sent

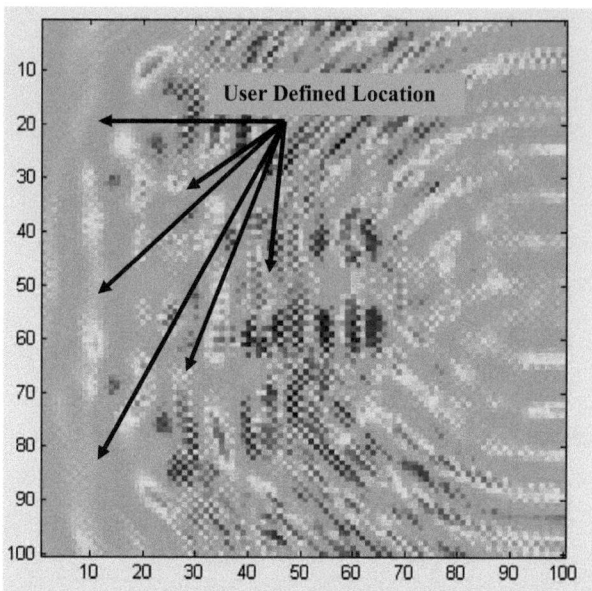

Figure.3.32. Screenshot Result for Retransmit Stage in User Defined Model

The propagation of the acoustics wave from the array of transducers to the transmitter is demonstrated in Figure.3.32. The waves may be spotted to the boundary and attained to the transmitter in user defined condition.

Figure.3.33 shows the screenshot result for received signal from the array for the user defined condition. By comparing with the user defined condition, the span of received signal is wider than the first two simulation result.

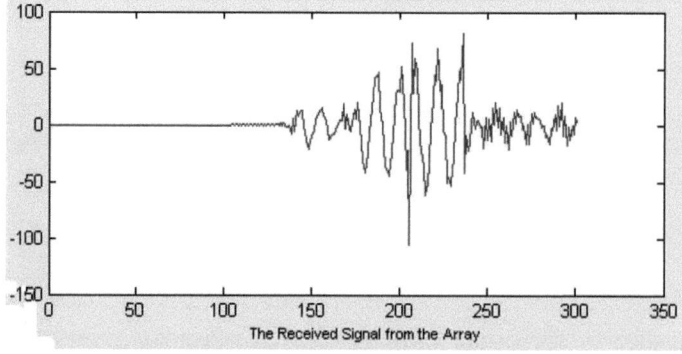

The Received Signal from the Array

Figure.3.33. Screenshot Result for Received Signal from the Array

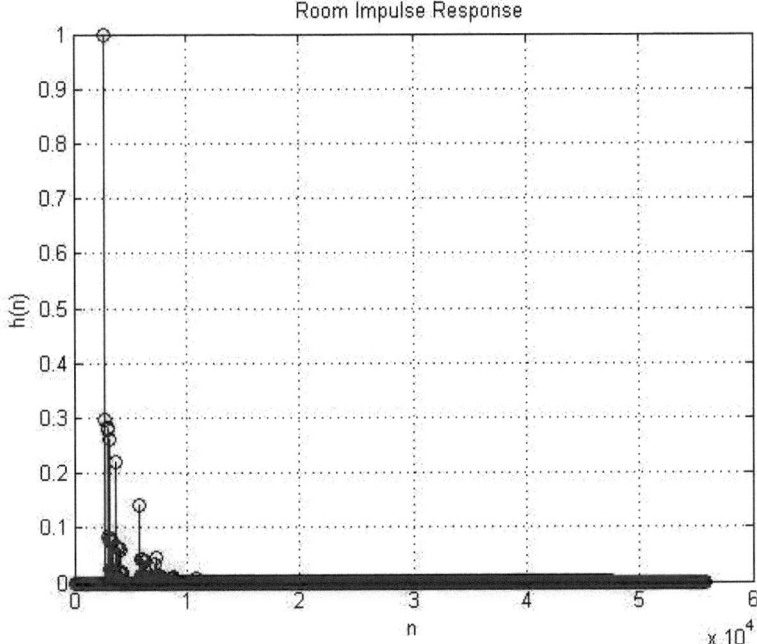

Figure.3.34. Room Impulse Response for Proposed Model

RIR is a program that calculates our room impulse response h(t). This program differs from the model in two ways. First it uses discrete time instead of continues time. Second, it finds $a_{i,j,k}$ and $e_{i,j,k}$ only when $a_{i,j,k}=1$. This is done to conserve memory. Figure.3.34 shows the room impulse response for proposed model. The room impulse response shows the satisfactory of the successful condition for proposed room in simulation study.

Chapter 4

Application Areas of Proposed Research

The beauty of time reversal signal processing is that one need not know any details of the channel. The step of sending a wave through the channel effectively measures it, and the retransmission step uses this data to focus the wave. Thus one doesn't have to solve the wave equation to optimize the system; one only needs to know that the medium is reciprocal. Time reversal is for that reason matched to applications with inhomogeneous media. An attractive aspect of time reversal signal processing is the fact that it composes use of multipath propagation. Numerous wireless communication systems are required to compensate and correct for multipath effects. Time reversal techniques employ multipath to their advantage by using the energy from all paths.

The effectiveness of time-reversal focusing is evaluated in the presence of a perfectly matched layer that changes the direction of the propagating waves, but does not incessantly scatter or block the propagating wave front. Interactions between the wave front and the surface layer are dependent on the depth and material properties of the asymmetric surface layer and its orientation in the medium with respect to the incident wave. Time reversal focusing is shown to perform significantly better than other excitation methods for the purpose of delivering energy to the location of source. While other focusing methods require some knowledge of the propagation medium characteristics such as propagation speed, time reversal does not require this information.

Time-reversal focusing is a powerful technique that allows propagating waves to be focused to a particular location. Time-reversal focusing is most useful when it is difficult or impossible to characterize the clutter and wave propagation speed in an area of examination. For this purpose, we use the auto focusing property of time reversal mirrors (TRM) to track in real time embedded system in its surrounding medium.

Chapter 5

Conclusion and Future Work

In optimizing the layout of the TRM for a time reversal experiment, it is originated empirically that for a simple source emitting a CW signal in a PML environment, the localization quality depends on the angular density of the TRM and the region of certainty. Time reversal focusing to exploit spatial/multipath diversity existing in rich scattering environments to improve the capability of target localization algorithms was introduced. Time reversal focusing was provided a built-in feature to adapt the transmitted waveform to the multipath environment and enhances the performance of the localization algorithms. The effectiveness of acoustics wave time-reversal focusing was observed in the presence of a perfectly matched layer. The surface layer changes the propagation velocity and direction of the acoustics waves and can manoeuvre them away from the location of a source. The surface layer further complicated the wave field since waves could propagate under as well as through the surface layer. Regardless of these effects, time-reversal focusing was effective and performed significantly superior than time-delay focusing.

We can extend to track the moving targets for time reversal focusing techniques by using UWB acoustics waves. Time-reversal techniques may also be extended to types of waves other than sound waves. Some researchers in the radar community are exploring their possible application to pulsed radar, using electromagnetic waves in the microwave range. Another type of wave occurs in quantum mechanics: the quantum wave functions that describe all matter.

References

[1] Amir Sulaiman, Kirill Poletkin and Andy W. H. Khong, "Source Localization in the Presence of Dispersion for Next Generation Touch Interface", 2010 International Conference on Cyber worlds, pp 82-86, 2010.

[2] D. J. Torrieri, "Statistical theory of passive location systems," IEEE Trans. Aerosp. Electron. Syst., vol. AES-20, pp. 183–198, Mar. 1984.

[3] T. T. Ha and R. C. Robertson, "Geostationary satellite navigation systems," IEEE Trans. Aerosp. Electron. Syst., vol. AES-23, pp. 247–254, Mar. 1987.

[4] A. H. Sayed, A. Tarighat, and N. Khajehnouri, "Network-based wireless location," IEEE Signal Process. Mag., vol. 22, pp. 24–40, Jul. 2005.

[5] T. Li, A. Ekpenyong, and Y.-F. Huang, "Source localization and tracking using distributed asynchronous sensors," IEEE Trans. Signal Process, vol. 54, no. 11, pp. 3991–4003, Oct. 2006.

[6] Tarik Yardibi, Jian Li, Petre Stoica, Ming Xue, and Arthur B. Baggeroer, "Source Localization and Sensing: A Nonparametric Iterative Adaptive Approach Based on Weighted Least Squares", IEEE Transactions On Aerospace And Electronic Systems Vol. 46, No. 1 January 2010.

[7] Rudolf Sprik, "Time-Reversed Experiments with Acoustics", NAG-Journaal nr. 174, maart 2005.

[8] Xueli Sheng, Fangfang Luo, Yong Guo, Longxiang Guo, "Research on Sleclective Passive Focusing Technology based on Dummy Iterative Time-Reversal Mirror", Science and Technology on Underwater Acoustic Laboratory Foundation, 2010.

[9] Blaine M. Harker, and Brian E. Anderson, "Optimization of the Array Mirror for Time Reversal Techniques Used In A Half-Space Environment", J. Acoust. Soc. Am. 133 (5), May 2013.

[10] Ibrahim El Baba, S´ebastien Lall´ech`ere and Pierre Bonnet, "Electromagnetic Time-Reversal for Reverberation Chamber applications using FDTD", ACTEA, July 15-17, 2009 Zouk Mosbeh, Lebanon, 2009.

[11] James R. Hopgood and Peter J. W Rayner, "A Probabilistic Framework For Subband Autoregressive Models Applied To Room Acoustics", Signal Processing Laboratory, IEEE Conference, 2001.

Printed by Books on Demand GmbH, Norderstedt / Germany